Jobs in
GREEN TRAVEL
and TOURISM

ROSEN
PUBLISHING®
New York

Jacqueline Ching

To Scotia and Dagny

Published in 2010 by The Rosen Publishing Group, Inc.
29 East 21st Street, New York, NY 10010

Copyright © 2010 by The Rosen Publishing Group, Inc.

First Edition

Library of Congress Cataloging-in-Publication Data

Ching, Jacqueline.
Jobs in green travel and tourism / Jacqueline Ching.
    p. cm.—(Green careers)
Includes bibliographical references and index.
ISBN 978-1-4358-3571-9 (library binding)
1. Ecotourism—Vocational guidance. 2. Green movement—Vocational guidance. 3. Green technology—Vocational guidance.
I. Title.
G156.5.E26C45 2010
910.23—dc22

                                                    2009016587

Manufactured in Malaysia
CPSIA Compliance Information: Batch #TW10YA: For Further Information contact Rosen Publishing, New York,
New York at 1-800-237-9932

On the cover: A scuba diver explores fire coral in the Red Sea.

On the title page: Left: An adventure guide leads the way to the glacial Uhuru Peak of Mount Kilimanjaro, in Kenya. Scientists predict that the glacier will disappear from the mountaintop by 2020 because of global warming. Right: A group of adventure tourists brave the rapids in a whitewater raft.

CONTENTS

Introduction

The future of tourism looks to be booming. Economists consider it to be the fastest-growing industry. Millions of travelers hit the road every day for business or leisure. People leave their mark wherever they go. On the ecological level, tourists and tourism can do a lot of damage. Travelers' globetrotting has made a significant impact on the environment, damaging natural resources, creating air and water pollution, and increasing the risk of global warming.

Since no one is about to give up traveling, people are seeking new ways to travel without causing ecological

damage. This has opened up a new field within the travel industry: green travel and tourism. Green travel used to mean "roughing it." It was designed for young, usually athletic people who wanted to get closer to nature by traveling to and exploring pristine natural areas like rain-forests, mountaintops, glaciers, and arctic tundra. The idea was that you took very little with you—just the bare essentials—and left behind little or no sign of you ever having been there. It ranged from simple hiking, climb-ing, or kayaking to more challenging sports, such as adventure cycling, ice climbing, or backcountry skiing—all while carrying everything that you needed to survive on your back.

Today, everyone from airlines and car rental agencies to hotels and resorts have an interest in becoming more "green." They are meeting the wants and needs of environmentally conscious consumers who are trying to lessen their negative impact on their destinations. There are many ways to do this. Travelers and tourists can start by learning about the culture and economy of the community they are visiting and try to support both in the most positive and sensitive ways possible. For example, green travelers must be vigilant about using the least possible amount of nonrenewable resources in their host country. At the same time, they must also try to spend their tourist dollars on local goods, made by local producers for locally owned stores and businesses. In this way, the community's economy is strengthened, and the money does not get diverted to big international corporations like those that own and operate fast food, coffee, and clothing chain stores. Airlines, hotels, and resorts have begun to realize that people want to travel, but not at the expense of the environment or local communities. Increasingly, tourists want to do the right thing, and they're looking for better choices.

Now that environmental protection is an increasing concern for both travel industry insiders and the tourists they seek to accommodate, existing jobs are evolving and new ones are being created. For example, hospitality managers were once primarily interested in keeping their guests comfortable, safe, and entertained. Now, however, they look for ways to achieve these goals while also promoting energy efficiency, water conservation, and other ecological efforts. And that's just the tip of the iceberg. In this book, we'll look at dozens of careers within the broad field of green travel and tourism. You will read about many concepts that are relatively new, such as carbon offsets and renewable energy projects. While the green travel and tourism industry is just in its infancy, many colleges and universities are preparing their students to work in this rapidly expanding field. The opportunities are great—and growing—for anyone interested in travel, adventure, hospitality, and saving the planet.

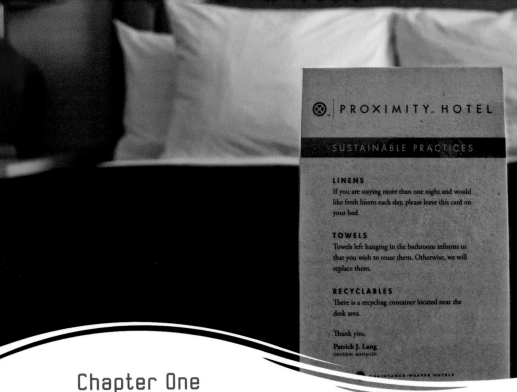

Chapter One

What Is Green Travel and Tourism?

The growth of green travel and tourism (also known as "ecotourism" or "sustainable tourism") reflects a change in people's attitudes toward, and relationship with, the environment. For thousands of years, travelers have visited historical landmarks and natural wonders. It is only in the last two decades, however, that people have become more aware that their presence makes an impact—often a

By some estimates, hotels use an average of 209 gallons (791 liters) of water each day for each occupied room. To reduce this water usage, more hotels now offer guests the option of reusing towels and sheets to limit the amount of laundry generated.

negative one—on the local surroundings. In fact, some forty-three million American travelers now describe themselves as "ecologically concerned," according to the U.S. Travel Data Center.

Ecotourism, as a concept and practice, began to emerge in the 1970s. But it first became widely familiar in the early 1990s. It is becoming increasingly popular as the public becomes more aware of the destructive—and avoidable—situations that are sometimes created by their travel activity. This, in turn, has created a growing need for experts in new areas of the travel industry, such as cultural heritage tourism and community-based tourism.

How do we define green travel and tourism? According to the International Ecotourism Society (TIES), the largest and oldest ecotourism organization in the world, the main principles of sustainable tourism are:

Ecological responsibility. Travelers should seek out accommodations and entertainment that do not harm the environment. One's lodging should engage in environmentally sound, low-carbon practices like water and energy conservation, recycling, and waste reduction.

Cultural sensitivity. Travelers should show respect for the different cultures that are being visited and fit in with and honor local customs, not demand that their hosts accommodate their cultural habits.

Supporting the local economy. Tour operators and their lodgings should employ local staff at a decent wage. Tourists should purchase their food and supplies locally and support local festivals and other events.

Experiential richness. Tourists should get involved in meaningful activities, ones that will educate them about the community and its culture, history, and traditions, and enrich the community, helping it thrive long after the travelers have left. For example, travelers might join a group that goes to a locality to help restore a historic building, dig much-needed wells, or build a school or medical clinic. The experience should be enriching for both the tourist and the community that he or she visits.

Types of Green Travel and Tourism

Before describing the types of jobs available in green travel and tourism, it would be useful to have an overview of the main categories or types of ecotourism.

Adventure Tourism

Before there was adventure tourism, there were adventure sports. Whitewater rafting, canyoneering, BASE jumping—these kinds of extreme sports pitted you against nature. At first, they appealed only to athletic types, offering them a unique way to challenge themselves while experiencing nature in a highly active, engaged way. Today, there are more of these types of activities, designed for a broader base of tourists who want to experience nature intensely but not necessarily in such a high-adrenaline, extremely physical sort of way. Some of these adventures can be contemplative, meditative, or spiritual, such as rides on the Trans-Siberian Railway, a retreat to a remote monastery, or walking tours of the rural regions in which writers like William Wordsworth and Thomas Hardy lived and worked.

Adventure tours can be found on any continent. Tourists can trek across the Himalayas or Kilimanjaro, dive into the Red Sea, cruise to Antarctica, or photograph cheetahs in Namibia, where the largest free-ranging population of these endangered wild cats are found. Unlike the average vacation, these activities offer tourists an added element of danger, along with once-in-a-lifetime access to sights that

At the Masai Mara Game Reserve in Kenya, eco-tourists have the adventure of a lifetime without harming the area's wildlife.

11

few people ever get to see. It falls on travel and hospitality specialists to manage these added dangers and to coordinate the tour experience so that safety remains paramount, while entertainment, enjoyment, and the spirit of adventure are preserved.

Coastal and Marine Tourism

Activities associated with coastal and marine tourism include cruises, boating, snorkeling, diving, swimming, and fishing. Coastal and marine tourism depends on the maintenance of clean and safe water for its continued success and popularity. Tourism industry officials must lobby local, state, and federal governments to put in place programs that ensure clean water; healthy coastal habitats; and a safe, secure, and enjoyable environment. It is also up to the tourism industry to help protect fish and other marine animals, wetlands, coral reefs, and other marine life that are a major draw for ecotourists. To achieve these protections, marine biology specialists are needed. Coastal destinations also have to manage the impact of weather hazards, such as hurricanes and tsunamis, calling for special expertise in things like hydrology, meteorology, and civil engineering.

Cultural Heritage Tourism

Cultural heritage tourism has always been at the heart of leisure travel for those who like to visit historic sites and cultural landmarks. Yet the term "cultural heritage tourism," first used in the 1970s, applies specifically to travelers

The ocean offers rare sights for more adventurous tourists. In return, tourists must help protect marine life and their habitats.

who want to gain a deeper understanding of a local culture. It defies the unfortunate stereotype of the clueless and obnoxious tourist. This is the traveler who visits a foreign country, snaps a few photos, complains about the local food, says everything is better where he or she came from, and leaves with nothing but cheap souvenirs (made somewhere other than the country being visited). People who seek a cultural heritage vacation want to fully experience the local culture, come to understand it, and leave it as untouched as possible.

Some trips that are offered give tourists a chance to meet, and sometimes live with, local people or tribes. Cultural heritage tourism also involves some cooperation with organizations that have emerged to safeguard disappearing elements of cultural heritage. This includes everything from languages and traditional practices to buildings and territory, such as traditional farmland, villages, or forest homes. Therefore, careers in the field of cultural heritage preservation can combine the twin interests of tourism and preservation.

Our Common Future

Some 113 countries agreed to an action plan at the Stockholm Conference in 1972. The plan laid out a framework for future environmental cooperation. This was reaffirmed in *Our Common Future*, a report published in 1987 by the United Nations. Since then, there have been increasing efforts to promote the theory and practice of sustainable tourism.

Community-Based Tourism

In community-based tourism, the local community is actively involved in planning and development. It is the opposite of the kind of tourism represented by large-scale hotel resort chains. This more grassroots and homegrown brand of tourism is developed in order to sustain and even rescue a local community. Innovative methods to attract tourists are used instead of large sums of investment dollars flowing in from outside of the community. The money that tourists spend stays within the local community—unlike in the hotel-and-resort model of tourism, where profits are often sent to corporate headquarters, rather than being invested back into the community.

Tourism typically has a negative impact on the poor residents of a tourist area. For example, these residents are often displaced for the sake of resort or theme park construction. In exchange for the encroachment on their homes, properties, and communities, they often participate in the tourism industry only as low-wage workers. Community-based tourism, however, seeks to empower the poor and tries to create a more positive social and economic impact. Many specialists are needed to make this happen, from those who are experts in microfinance to those who can educate and train local residents in artisan crafts, hospitality services, and other tourist-friendly enterprises.

The promise of job creation in the green travel and tourism industry is great, especially as more and more governments and nongovernmental organizations (NGOs) begin investing in sustainable tourism. Belize, a Central

Tourists can make a positive social and economic impact when they buy souvenirs that are made locally and that profit the local community.

American country with three indigenous Mayan populations and natural treasures like the Belize Barrier Reef, recently invested $13.5 million in a four-year sustainable tourism plan. NGOs—including the International Centre for Responsible Tourism (ICRT), a network of organizations in Britain, Belize, Canada, Germany, India, South Africa, and West Africa—also promote and develop sustainable tourism. This growing activity, interest, and investment translate into more job opportunities for those committed to a career in green travel and tourism.

Chapter Two

Jobs in Green Transportation

One of the biggest challenges to the environment is the burning of fossil fuels, such as oil, coal, and natural gas. These are nonrenewable resources. Fossil fuels take millions of years to form. At the rate at which humans burn them for energy, they will soon be gone. Additionally, when burned, fossil fuels produce air pollutants—lead, carbon monoxide, nitrous oxide, and chlorofluorocarbons, to name just a few.

Traditional methods of transportation, including fossil fuel–burning automobiles, are responsible for emitting several million tons of pollutants into the atmosphere each year. Hopefully, a shift to electric-powered vehicles, such as this one, will lessen global warming.

The burning of fossil fuels releases carbon dioxide. Increased levels of carbon dioxide emissions are a serious environmental problem, as this "greenhouse gas" causes heat from the sun to be trapped in Earth's atmosphere. Scientists fear that the resulting warming of the planet has caused long-term changes in Earth's climate. These changes may lead to catastrophic glacial retreat, a rise in sea levels, extreme weather conditions, habitat destruction, and mass species extinctions. In fact, scientists now believe that the pace of climate change is faster than previously thought. A 2001 United Nations report warned that between 1990 and 2100, average global temperatures would rise between 2.5 and 10.4 degrees Fahrenheit (1.4–5.8 degrees Celsius), an increase that would greatly stress and perhaps kill many forms of plant and animal life.

Green travel has the same goals as all other environmental conservation efforts. One of these is to limit the use of fossil fuels. One of the most urgent issues within the green travel industry is transportation. The activities of the transportation industry (which include travel by car, boat, plane, and train) emit several million tons of pollutants into the atmosphere each year. It is estimated that a family of four flying across the Atlantic Ocean from North America to Europe contributes more emissions during the flight than their entire domestic energy use in a year, according to John Stewart in *Red Pepper*, a political advocacy magazine.

Yet in order to travel, people must have a way to reach their destination, and, once there, move around to see the sights. Today, travelers have a growing number of options, many of them being green, or at least greener, alternatives to high-carbon-emitting transportation. For example, energy can be saved and emissions reduced if people use a hotel carpool van or an electric car instead of a rental car or taxi. But all of these greener travel options have to be developed, promoted, and managed effectively if tourists are going to become aware of them and choose them.

Carbon Offset Managers

At present, it is not possible to fly an airplane without burning fossil fuels. In addition, most people still drive fossil fuel–powered vehicles. But a growing trend is aimed at reversing the damage caused by these transportation activities. More and more companies in the transport industry, such as Expedia, Travelocity, REI, and Continental Airlines, are creating carbon offset programs.

A carbon offset is a way to represent a reduction in greenhouse gas emissions. It's also a way of managing one's carbon footprint by contributing to renewable energy projects, such as planting trees and installing solar panels. The goal of environmental efforts like carbon offsets is to be carbon neutral. This means that for every activity that releases carbon to the atmosphere, the same amount of carbon should go unconsumed through another, greener activity.

One carbon offset represents the reduction of a metric ton of carbon dioxide, or its equivalent in other greenhouse gases. In the case of Continental Airlines, customers can calculate the carbon footprint of their itinerary, translate it into a dollar amount, and then make a contribution to Sustainable Travel International (STI), which would offset the emissions produced by those flights. STI, a nonprofit organization, uses the donations it receives from carbon offset programs to pay for renewable energy projects, such as building a hydroelectric plant in Indonesia. This plant generates electricity through the power of rushing water, rather than the burning of fossil fuels.

Carbon offset managers manage their company's offset programs. They have to identify credible green organizations to work with; these are the organizations that will receive the company's offset donations and funnel the money into green projects. This takes some research. In the United States, there are more than six hundred organizations that develop, market, or sell offsets. But there isn't a single regulatory body to oversee them and set clear standards and regulations.

Carbon offset managers will need a college degree, preferably in business or economics, because they have to be able to analyze the companies they hope to partner with and ensure that they are financially sound and ethical. Knowledge of several foreign languages is also a plus, as this is a highly international field. The salary starts at about $75,000, plus benefits, depending on the person's experience with carbon reduction projects.

Carbon Surveyor/Assessor

A carbon offset manager works for businesses, such as airlines and hotels, which offer their customers the opportunity to buy carbon offsets to reduce the impact of their fossil fuel–based activities, such as air travel

The Kyoto Protocol

This poster marked the tenth anniversary of the Kyoto Protocol in December 2007.

Adopted in December 1997 in Kyoto, Japan, the Kyoto Protocol is an agreement by industrialized nations to reduce their collective greenhouse gas emissions by 5.2 percent below 1990 levels. It came into force on February 16, 2005. Many scientists believe that a reduction of 5.2 percent will do little to slow down climate change. Nevertheless, the Kyoto Protocol is an important first step in international cooperation on an extremely urgent global issue.

and electricity use. A carbon surveyor, on the other hand, works for carbon management businesses. These are the companies that help other companies determine their carbon emissions and energy and water usage. They then consult with their clients to show how energy reductions can be made while also offering them the opportunity to offset carbon emissions by contributing to green projects.

A carbon surveyor/assessor studies the business operations of his or her clients, writes an energy audit report, and recommends emission reduction strategies. The assessor presents clients with carbon offset programs that they can participate in to help them conform to local, national, and/ or international pollution and energy use laws and standards. A degree in engineering is required. Knowledge of the energy market and renewable technologies, as well as several years of on-site energy surveying, are also a must. Salaries start at about $50,000 a year, plus benefits.

Environmental Health Project Manager

Airlines and airports are working on making environmental health a standard practice. The responsibility for this often falls on an environmental health project manager. All airlines operating in the United States must employ a green operations program because of safety standards enacted by the U.S. Environmental Protection Agency (EPA). These standards require the proper disposal of oil, fuel, sealants, and other harmful chemicals associated with air travel and plane maintenance.

There is also growing governmental and consumer pressure on airlines to employ greener in-flight operations. For example, Delta, American, and Continental are three of the leading airlines that have recycling programs. During a nine-month period in 2008, 500,000 pounds (226,796 kilograms) of aluminum, paper, and plastic from Delta flights were recycled. American Airlines also participates in a battery recycling program. By recycling, airlines not only save money and energy but also avoid the greenhouse gas emissions caused by producing more of the same products. Yet challenges remain. If the airport where a plane lands doesn't have a recycling facility, onboard trash will end up in a landfill. As it is, "[T]he U.S. airline industry discards enough aluminum cans each year to build fifty-eight Boeing 747 planes," states the Natural Resources Defense Council.

Managers of environmental programs manage in-flight recycling programs. They create additional programs and procedures for anything that can be recycled, from paper cups, magazines, water bottles, and soda cans to airplane carpeting, upholstery, and engine oil. To prepare for a career as an environmental health project manager, you should earn at least a bachelor's degree. You should have experience and training in safety, industrial hygiene, risk management, and sanitation, including a food management and waste management degree or certificate programs. Salaries range from $50,000 to $60,000 but increase with experience.

Research

Researchers are needed in the green transportation field. Research and development is being done to build cars and airplanes that operate on renewable energy. For example, in early 2009, Continental Airlines rolled out its Boeing 747-800, an aircraft powered by sustainable biofuel. The biofuel blend is made from such sustainable sources as algae and jatropha plants. These do not impact food crops or water resources or contribute to deforestation. In addition, automakers are currently rushing to develop hybrid and fully electric cars that will require far less, if any, fossil fuels in order to operate.

Dr. Robert Manurung of Indonesia sits next to a sample of jatropha oil that was used to power the first fully plant-fueled car. It completed a 2,000-mile (3,129 km) road trip.

One of the most important tasks of a product researcher and developer in the aviation and automotive industries is to create and test new, renewable, and clean-burning fuels and lubricants for use in automobiles and airplanes. Tasks like these require at least a bachelor's degree in chemistry, biology, or engineering. Some schools offer bachelor's programs in renewable energy. These include the Oregon Institute of Technology, Illinois State University, Appalachian State University, and State University of New York in Canton. Starting salaries range from $45,000 to $60,000 for candidates with relevant degrees from a four-year program.

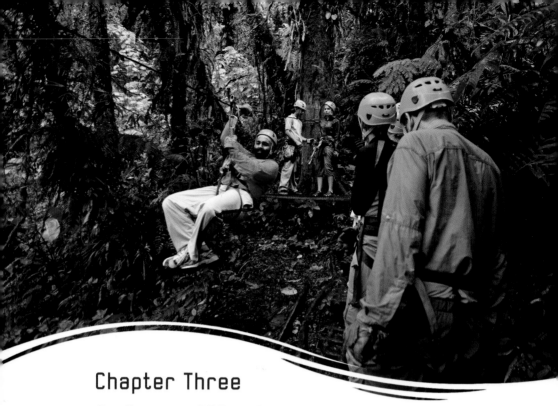

Chapter Three
Jobs with Green Tour Operators

One of the biggest challenges of sustainable tourism is certifying green tour operators. The industry is simply so large that it is difficult to set standards and certification schemes that are accepted worldwide. There are more than six hundred tour operators in North America alone. Many of these companies are small, and their few employees must multitask. They may perform several jobs in a single day, including clerical work (filing, record keeping, and managing correspondence), marketing, advertising,

Tour operators help tourists enjoy once-in-a-lifetime experiences that do not harm the environment, such as sailing a jungle canopy on a zip-line.

and the actual leading of tours. Most of the jobs available are in tour management or tour guiding, and many of these jobs are seasonal. They are not year-round, full-time jobs. If seasonal work is what you are looking for, make sure to research job application deadlines. In the Northern Hemisphere, there are far more jobs available in summer months than in winter months.

No tour operator will require job applicants to have a degree from a guide training school. However, such programs can help you with job placement and making contacts with industry leaders. In addition, coursework in subjects like ecology and environmental science will set you apart from other applicants. Knowledge of a foreign language is also an advantage if your destinations are overseas.

Tour operators often give their new employees an introductory course that includes information about their destinations, as well as an overview of company policies and procedures. Those tour operators that are specifically "green" will take extra steps to make sure their employees understand and are able to fulfill the company's environmental goals.

Tour Manager

Tour managers oversee all aspects of the actual tours. These include finding eco-friendly accommodations for travelers, arranging for low-carbon transport, training bus drivers and guides, and the careful scheduling of each day's tour activities. It's a job that can have an unpredictable schedule. You might find that you are working every day during the tourist season and then are idle for several months. And during high season, it's potentially a twenty-four-hour workday, seven days a week, because you might be entertaining guests far into the night and have to deal with emergencies at all hours. As a tour manager, you are always on call to respond to the needs of your guests and the tour guides you are managing.

Tour managers are responsible for every detail of the trip. Things inevitably don't go exactly as planned. Unforeseen challenges do arise. As a result, tour managers have to be good problem-solvers. And they must be

The International Tour Management Institute, at http://www. itmitourtraining.com, was the first certified program to train tour directors and travel guides. Since then, more programs have emerged to meet the demand for trained travel professionals.

able to calm anxious or disappointed guests and ensure that everyone has a positive experience, regardless of any mishaps that occur.

In order to set himself or herself apart from the competition, a green tour manager should take traditional and green tourism courses and earn a relevant degree in archaeology, geography, history, and/or business with a foreign language. Tour managers may periodically receive further training on the job to learn about new destinations and cultures. To become even more attractive to a prospective employer, a person can now take a written and oral exam to earn a certificate of tour management accreditation, given by the International Association of Tour Managers (IATM). The International Guide Academy (IGA) offers short courses than run from ten days to sixteen days. They have even offered courses aboard cruise ships. At the end of an IATM course, you receive a certificate and lifetime job placement assistance. In addition, a company called Professional Tour Management Training offers courses at colleges and universities in California, as well as online. On average, tour managers can earn from $2,500 to $3,500 for a tour that lasts twelve to fourteen days. This includes salary and tips.

Tour Guides

Being a green tour guide and traveling to out-of-the-way and exotic vacation destinations can be exciting. "The tour guide has to love people, love to talk, and love research," says Daniel Slater, president of the IGA, one of the leading educational institutions in the field of travel and tourism. No prior experience is necessary, but you have to be a people person and be able to handle a wide range of personalities, especially the difficult and demanding ones.

A green tour guide should be a local expert of sorts and serve as the intermediary between guests and the host community. To fulfill this role well, you have to do research on the local people and their customs, culture, and history, as well as the region's flora and fauna (plants and animals). This knowledge and expertise will help ensure that both your guests and your hosts benefit from the tour experience. If you don't help educate your guests about the community and encourage them to fully enter into its culture, misunderstandings may arise and your hosts may begin to feel resentful of the tourists' presence. Some tour operators require their tour guides to be licensed. Usually, guides will have to take a written test about their area of expertise. Although no prior guide experience is necessary, outfitters, tour operators, and independent companies offer their new guides short training courses.

Tour guides can earn from $25 to $100 a day, depending on the type of tour and other factors. They can either receive a salary (plus benefits), or they can be paid by the day. Tour guides who are paid by the day receive tips from their customers, which can increase their income. Some guides make as much as $50,000 a year. A tour guide's out-of-pocket expenses are minimal, since food and lodging are often provided. The type of destination and clientele can affect how much a tour guide earns. It can be a challenge to make a living wage throughout the year because the work is often available

Tourists must be led through the Los Glaciares National Park in Argentina by trained guides. The park, a popular destination for international tourists, has been declared a Natural World Heritage Site by the United Nations Educational, Scientific, and Cultural Organization (UNESCO).

only during the tourist season (depending on the location, this is usually about a six-month period). If your tour operator offers international tours in both the Northern and Southern Hemispheres, you can try to secure guide jobs in each hemisphere's respective tourist season, thereby maintaining year-round employment.

Adventure Guide

An adventure guide takes a small group of people on trips that can last a day, a weekend, a week, or even longer. These trips usually involve different outdoor sports, from rafting and hiking to climbing and caving, so physical fitness and athletic ability are definite requirements. Adventure tours give you a chance to see some of the world's most beautiful, remote, and unspoiled places. One of the most important aspects of your job is to ensure that your party leaves the places it visits as pristine as they were before your arrival. This is known as a "Leave No Trace" ethic. It requires guides to adopt various strategies and precautions. These include only lighting fires in appropriate places at appropriate times, planning meals carefully and bringing the proper equipment to minimize trash, and using existing trails and designated campsites to avoid trampling on vegetation. You should also use biodegradable soap, properly dispose of wastewater far enough away from freshwater sources, and carry out everything you carried in, including empty food packaging.

Other guide duties include some office work; maintaining and renting out sports equipment and outdoor gear; taking customer reservations; and making any necessary

Outward Bound

Outward Bound is a nonprofit company that organizes wilderness-based programs across the United States. Some of its courses are geared toward students and struggling youths. Adventures, which range from sailing in Maine to rock climbing in Appalachia, can be physically grueling. Some trips last a few days, while some last a few months. Each year, Outward Bound offers educational scholarships to students—in 2009, the total scholarship amount was $2 million. Hundreds of high schools and colleges grant academic credit for Outward Bound expeditions.

reservations with campsites, transportation companies, and parks. Many outfitters and other outdoor specialists offer guide training. Although no experience is necessary, outfitters, tour operators, wilderness survival specialists, and independent companies offer short survival and outdoor guide training courses. You can take one of these courses in the spring and be certified in time for the summer season. The pay for adventure guides is similar to that for tour guides.

Raft Guides

Many states require raft guides to be certified by the state rafting guide association. To be certified, you will need to complete a course or take a written test. The job of a raft guide is similar to that of an adventure guide. Important parts of the job are instructing your customers on how to

Raft guides make sure tourists have a thrilling but safe trip by instructing them on safety precautions and what to expect on the water.

properly handle themselves in the boat, educating them on what to expect on the water and when to expect it, and being knowledgeable about safety precautions and emergency response. Like other guide jobs, it helps if you can provide vital information in an entertaining, supportive, and friendly way. You will also have to test, maintain, and repair your rafting equipment.

Rafting guide training courses are available. For example, the Colorado State Parks and Outdoor Recreation Department offers an introductory course for $395. It covers, among other things, different paddle strokes, navigation, and safety skills. You must also become certified in first-aid techniques. Raft guides earn about $30 per rafting trip. In addition, they receive about $15 to $20 in tips per trip. Guides can give as many as five or six trips per day, which means they can make from $270 to $300 per day. But other guides and tour operators prefer to give fewer, more personalized trips that can last longer or even take place over several days.

Nonguide Jobs

Adventure tour companies hire cooks, base camp workers, and drivers—often without previous experience. There is also a need for customer service representatives (to take reservations, book group tours, send out brochures, and answer questions), sales representatives, and Web site administrators and designers. Working in these jobs over one summer can get your foot in the door and give you a chance to get hired as a guide the next season. There are no particular educational requirements for these kinds of jobs, but previous relevant work experience is recommended. These jobs usually pay an hourly or weekly wage.

Chapter Four

Jobs in Public Parks and Forests

Governments are creating policies to encourage eco-friendly practices in tourism. They are also the employment source for thousands of green jobs. For example, in February 2009, U.S. president Barack Obama signed into law an economic stimulus package that includes $920 million for national parks. This will create jobs as well as help preserve America's heritage and natural treasures. This is work that is much needed.

The Statue of Liberty National Monument is managed by the National Park Service. U.S. park rangers guard the area and make sure everyone behaves responsibly.

There will be construction jobs at places like Acadia National Park in Maine and Death Valley National Park in California, where roads and trails need repair. There will be jobs at Valley Forge National Historical Park in Pennsylvania, where historic buildings need to be restored. From building and trail maintenance to management positions, careers will be built in the National Park Service, the U.S. Forest Service, and other state and federal agencies.

The four major federal public land managers are the U.S. Forest Service (USFS), the National Park Service (NPS), the U.S. Fish and Wildlife Service (USFWS), and the Bureau of Land Management (BLM). In addition to the national parks, there are 155 national forests, managed by the USFS; 548 national wildlife refuges, managed by the USFWS; 27 million acres (11 million hectares) of land that make up the National Landscape Conservation System, managed by the BLM; forty-seven biosphere reserves; and a growing number of marine protected areas.

Volunteering is a good way to get your first job in land management. Being a volunteer will allow you to make contacts, form relationships with superiors, and be in a good position when a paying job opens up. Volunteers can be involved in many aspects of park management, from orienting visitors or leading field trips, to maintaining the park or working on studies that can help further park development.

Park Superintendent

The park superintendent is usually responsible for the management of the state or national park and all its recreation services. He or she oversees the smooth operation of camping, lodging, and dining facilities, as well as the preservation of natural assets. Other responsibilities include the maintenance of buildings and grounds; managing sanitation, public relations, and public safety; making sure state and federal regulations are followed; and account keeping and budgeting. Park superintendents work for the National Park Service or for a state's parks department, but their skills can be transferred to jobs in private eco-parks, wildlife sanctuaries, and other private nature preserves.

A written exam is often required as part of the job application process. It is the same exam given to job candidates for park ranger positions. A bachelor's degree in natural resource management, conservation, forestry, botany, zoology, geology, natural history, business administration, or public administration is recommended, as is at least one year of related work experience. Salary ranges are as

The U.S. Forest Service

The U.S. Forest Service manages the nation's 155 national forests and twenty national grasslands. It falls under the U.S. Department of Agriculture. Its law enforcement unit, the U.S. Forest Service Law Enforcement and Investigations (LEI), makes sure federal laws are not broken on national forestlands. It has officers that patrol and carry firearms, special agents who perform investigations, and K-9 units and mounted police.

broad as $24,000 to $36,000 and $75,000 to $130,000. These ranges reflect the difference in employer, location, and the candidate's experience.

Forest/Wilderness Rangers

Rangers perform the well-known tasks of greeting visitors to parks and forests, and running education programs for guests. Their lesser-known duties include measuring emissions from cruise ships, studying harbor seals, or tracking animals to see if any act strangely and may be sick.

A U.S. forest ranger patrols Baldy Mountain, West Virginia, on an all-terrain vehicle (ATV) after a brush fire.

Sometimes, rangers will help search for a hunter, camper, hiker, or child who may have gotten lost. A ranger's day starts early, and it is always unpredictable.

Forest rangers work for the U.S. Forest Service or a state forestry service. They wear uniforms and are required to carry firearms. They sometimes have to use them. Therefore, one of the requirements to become a ranger is to pass a gun safety or firearm training class. Candidates must have a bachelor's degree in environmental science or a related subject. Or they must have an associate's degree in forestry, forest management, environmental science, or a related subject. They must also take and pass a state civil service test. There may also be a vision and physical fitness test. Depending on different factors, such as location and experience, a new ranger can earn $40 an hour or $20,000 to $30,000 a year with benefits. An experienced forest ranger can see his or her salary increase to $75,000 to $130,000.

Trail Coordinator

A trail coordinator develops and manages the trails in a national or state park or forest. Maintenance of trails involves management of natural, historical, and cultural resources and the careful balancing of biking, snowmobiling, and other recreation issues. If planned right, such trails help conserve the local environment while allowing visitors the opportunity to get out and enjoy physical activity in nature. Trail coordinators give tours to the public. They also have to manage staff and work with government agencies on issues relating to the budget, park construction, and

Each of the hundreds of thousands of trails at parks around the country have to be maintained. At this California state park, a maintenance worker pounds the ground flat to make for an easier hike.

the acquisition of additional park property. A bachelor's degree in resource management is required.

The position of trail coordinator can pay an hourly wage, but it is often a job that is performed by volunteers. As such, it does offer a great entry into the park or forest system and an opportunity to move up the ranks into paying positions.

Research Ecologist

A research ecologist studies the effects of tourism on local fauna and flora. Green destinations have to protect their

Occasionally, animals that live at tourist destinations, such as this 9-foot-long (2.74 meters) alligator, have to be moved for their own safety. Sometimes, biologists pick them up for monitoring or other research purposes.

assets by continued monitoring of the interactions that take place every day between people, animals, and plants in parks, forests, and other sensitive ecosystems that are popular with tourists. If human interaction is found to pose any danger to the natural habitat, or if wildlife is somehow becoming a threat to visitors, the research ecologist will ring the alarm bells. They will then study the matter and propose ways to protect humans and wildlife alike, all while preserving the existing habitat and ecosystem.

This is a job for a highly trained scientist, with a degree in chemistry, ecology, biology, or microbiology. Starting salaries run from $30,000 to $50,000. Higher salaries may be earned if you work for a private environmental consultancy or are a corporate ecologist.

Chapter Five

Jobs in Green Hospitality

Hotels and resorts have good reasons to "go green." Given our dwindling supplies of fossil fuels and the onset of global warming, green building and operating practices represent the future of the hospitality industry. Furthermore, because of the growing interest among consumers for green options, going green will increase the revenue of hotels and resorts in the long run. More and more businesses are being rated for their "eco-friendliness,"

As more green hotels are built, like this one in Greensboro, North Carolina, tourists have better and more eco-friendly travel options. This hotel uses state-of-the-art energy-saving devices to heat water and filter the air.

allowing consumers to make the best possible choices when planning what products to buy, what places to visit, and where to stay. There is media coverage of the best and worst eco-friendly destinations. For example, *U.S. News & World Report* recently published a list of "the best (and worst) eco-friendly ski resorts."

What are green hotels? They are hotels that use water-saving measures and energy-reduction strategies, and they minimize and recycle solid waste. There are many ways that hotels and resorts can go green. For instance, they can use compact fluorescent bulbs instead of traditional incandescent lightbulbs, use biodegradable cleansers and detergents, and install solar panels and "green roofs." (Green roofs are landscaped roofs that provide natural, nontoxic drainage and insulation). Hotels can also install ultraefficient, dual-flush toilets and rainwater-harvesting tanks to catch rainwater for the flushing of these toilets. These hotels and resorts need people to implement these green initiatives, install green equipment and systems, and manage and maintain them. Once these initiatives are in place, the results can be dramatic: a Hyatt in Chicago was able to reduce its waste by 80 percent.

Builder/Retrofitter

The building of the future has arrived. It has solar panels, under-floor heating powered by geothermal heat pumps, electronic sensors that measure the outside temperature and adjust the room temperature accordingly. It has a rainwater harvesting tank that catches water for flushing toilets and running the dishwasher and laundry machine, a super-insulated exterior glazing that minimizes the building's loss of heat, and so on.

Very soon, it won't be enough for eco-friendly hotels, resorts, and other businesses to adopt just a few token green technologies and practices for public relations purposes. Increasingly, consumers are demanding carbon-neutral buildings built from the ground up. Green builders and retrofitters will be needed to meet this growing demand to build thoroughly eco-friendly buildings from scratch and to adapt existing buildings to be more energy efficient.

Knowledge of construction basics is needed for a builder and retrofitter. This includes carpentry, electrical, plumbing, or HVAC (heating, ventilating, and air-conditioning) skills. Trade schools, technical schools, and community colleges all offer training and certification in these trades. Some of these schools are already emphasizing green building techniques and technologies. Salaries vary widely, depending on whether you work for yourself, for a contractor specializing in green building, or for a green construction firm. But the starting pay tends to be about $30,000 a year.

Sustainable Design Architect

The architect and builder work hand in hand. The architect draws up the plans for the building, and the builder makes them a concrete reality. A sustainable design architect does all the same things that a traditional architect does: preparing designs and getting them approved by the city or township, supervising construction, and working with the builder to troubleshoot any problems that crop up.

Built in 2004, the University of Texas School of Nursing building is one of the largest and most sophisticated green academic buildings in the Southwest.

A sustainable design architect also has specialized knowledge of how to meet green standards. Throughout the United States and in thirty countries around the world, the main standard that is used is called LEED, or Leadership in Energy and Environmental Design. It was developed by the U.S. Green Building Design. LEED is a third-party certification program and the nationally accepted benchmark for the design, construction, and operation of high-performance green buildings. LEED gives building owners and operators the tools they need to have an immediate and measurable impact on their buildings' performance. LEED promotes a whole-building approach to sustainability by recognizing performance in five key areas of human and environmental health. These are sustainable site development, water savings, energy efficiency, materials selection, and indoor environmental quality. It provides standards for new construction; existing buildings; a building's core, shell, and interiors; schools, retail, and health care establishments; homes; and neighborhood development.

Aspiring architects can enroll in a five-year master of architecture program. Or you can get a (four-year) bachelor's degree in another field, such as mathematics, art, or engineering. Then you can continue on to a two-to-three-year master's program in architecture. Starting salaries range from $50,000 to $65,000, but there is great potential for professional growth and a larger income.

Plumbing Retrofitter

A plumbing retrofitter analyzes a building's existing plumbing and water usage, and then makes recommendations

Plumbers interested in retrofitting homes to be more "green" can use devices such as this rain barrel. It captures rainwater from the gutters for nondrinking and bathing uses in the home. This reduces the house's use of public water supplies.

on how to reduce water usage and improve the system's efficiency. He or she can install ultraefficient toilets as well as irrigation systems, water meters, and other conservation technologies.

Licensing requirements for plumbers vary from state to state. To get a plumbing license, you generally need to be trained as a plumber in a technical or trade school or a community college. You may have to pass a certification exam. Often, local community and technical colleges offer plumber certification programs. Training can occur on the job in apprentice-type arrangements. Plumbers generally earn about $70 to $90 an hour, plus benefits.

Green Event Planner

Important facets of the hospitality industry, both traditional and green, are industry conventions, business meetings, and other professional events. These special events allow large numbers of people to meet at the same location. These people are often in the same industry. Perhaps they even work together over the phone and Internet, but don't usually meet face-to-face. Other special events include private parties, such as weddings, proms, or bar/bat mitzvahs.

Green event planners are specialists who know where the best local restaurants, entertainment, and tourist spots are. They can smooth the path for corporate guests who are holding their event in a place with which they are unfamiliar. The green event planner has a lot of knowledge about the surrounding area and the hotel's green practices. For instance, he or she can point guests toward

Getting an Education in Green Building

These are just some of the universities that offer programs in green building:

California Polytechnic State University, College of Architecture and Environmental Design, in San Luis Obispo. Degree programs include bachelor and master of science in architecture.

Carnegie Mellon, College of Architecture, in Pittsburgh. It offers a twelve-month master of science degree in sustainable design.

Colorado State University, Continuing Education, in Denver and Fort Collins. This institution offers a green building certificate.

Frank Lloyd Wright School, in Spring Green, Wisconsin, and Scottsdale, Arizona. This school offers bachelor of architectural science and master of architecture degrees.

Georgia Institute of Technology, in Atlanta. Undergraduate and graduate courses are available at this institute.

University of Massachusetts, in Amherst. This school offers a building energy efficiency program (BEEP).

the locality's natural treasures or can recommend an interesting community event to attend or an organic local-foods restaurant to visit. The event planner will know if a hotel provides room discounts or free valet parking service to visitors who rent hybrid cars. When it comes to an eco-friendly event, a green event planner can give companies the reassurance that their event will be handled in a responsible way, including the recycling of all food packaging and food service items, the sending of left-over food to an area food pantry or soup kitchen, and the serving of locally grown foods that did not require a high carbon output to be delivered to the table.

College degrees are available for the event planner, although they are not a requirement for a successful career. If you come from a business background or an environmental sciences background, you can take supplemental classes and receive certification. It's useful to have knowledge of accounting, sales, and management. There is huge potential for income growth in this career. It is possible to make six-figure salaries, but expect to start at about $45,000 a year.

Chapter Six
Jobs in the Green Restaurant and Food Industry

At the mention of green dining, you may think we're talking about eating healthier and loading up on fruits and vegetables. That has something to do with it. Eating organic food grown without pesticides, chemical fertilizers, irradiation, and genetic engineering is not only better for the environment, but it's also better for

Eating organically grown food is better for our bodies and for the environment. Farmers and restaurateurs can help promote the importance of sustainable agriculture. This farmer holds handfuls of organically composted soil.

you. Plus, it's a question of choosing to eat locally. Much carbon-based fuel is emitted into the atmosphere to ship food products hundreds, sometimes thousands of miles from distant farms to your table or the corner restaurant. Yet a movement is afoot to encourage the consumption of locally grown foods in order to reduce that enormous carbon footprint. Indeed, green dining is as important an issue to the preservation of the planet as green transportation is.

The Green Restaurant Association (GRA), a nonprofit organization, has compiled a list of environmental guidelines for the restaurant industry. These include the use of locally grown foods and the use of recycled, tree-free, biodegradable, chlorine-free, and organic kitchen and dining supplies and food packaging. The GRA recommends nontoxic cleaning supplies; composting and recycling; and appliances that are energy-efficient and water-efficient.

World Summit on Sustainable Development

Since the Earth Summit in 1992, there have been periodic efforts to bring together world leaders to agree on business codes of conduct that will protect the environment, or at least reduce the harm being done to it. In 2002, the United Nations assembled and hosted the World Summit on Sustainable Development in Johannesburg, South Africa. The main problem was that the United States wasn't there. President George W. Bush boycotted it. Without U.S. support and participation, the summit's resolutions had little impact.

The GRA also provides a simple way of rating and certifying green restaurants. There are separate standards for newly constructed restaurants, retrofitted restaurants, and special events. For example, a restaurant gets three points for using ultraefficient toilets, five points for using hormone-and-antibiotic-free meats, and thirty points for being vegetarian. Establishments are also awarded a system of stars. For instance, a restaurant that has accumulated 470 points receives four stars.

Green Food and Beverage Manager

In addition to hiring, training, and supervising workers, the food and beverage manager plays the important role of purchasing food and drink supplies, dealing with vendors, keeping track of revenues and expenses, and making sure the establishment practices healthy and safe food preparation. He or she may be employed by restaurants, bars, hotels, resorts, and anywhere food and drinks are served.

Today's food and beverage manager has to be able to incorporate principles of sustainability. He or she must find out where to get supplies that are eco-friendly. For example, the food and beverage manager would research and test beverages like Park City Ice Water that are considered eco-friendly. Park City's packaging requires 75 percent less energy to produce and is recyclable. Park City's Ice Water flows directly into the bottling plant (in Sandy, Utah) from a glacier lying below the city. As a result, there is no need to transport the product in tanker trucks, which saves on gas, the burning of fossil fuels, and the emitting of carbon dioxide into the atmosphere. Other companies are beginning to offer their beverages in bio-degradable and/or compostable bottles, some of which are made from cornstarch, a renewable resource. It would be the food and beverage manager's job to find out about these products, contact their distributor, and negotiate a deal to carry them.

Want to become a food and beverage manager? A degree in a business or tourism-related subject is help-ful, as are several years of experience in the restaurant, bar, food service, or hospitality industries. The average hourly wage is $14 to $16, not including tips. The aver-age salary ranges from $40,000 to $55,000, not including tips.

Executive Chef

On television at least, the job of a top chef looks like a glorious position. But it actually demands serious commit-ment and hard work. It takes a lot of time and training to work your way to the top, and competition for the rela-tively few openings is fierce. Executive chefs oversee the kitchen and often do little of the cooking, although they do supervise and give instructions—right down to the place-ment of the garnish. They train cooks and select the menu, which, in a green establishment, means they have great influence on the restaurant's carbon footprint. A lot of time is spent outside the kitchen, doing research on trends in the restaurant industry and looking for organic and local growers and vendors. Other responsibilities include bud-geting, financial planning, and business development.

Many restaurants today ask their chefs to use produce grown locally and meat that is free of growth-stimulating hormones and antibiotics. These red bliss potatoes are organically grown.

The executive chef knows that customers who choose to patronize a green restaurant want their food to be organic and locally grown and produced. "Local" means the food is fresher, and fewer greenhouse gases were emitted transporting it to the market and then the restaurant. The executive chef knows where to buy organic, Earth-friendly ingredients that fit into the budget (since many of these foods can be more expensive).

If you wish to become a chef who specializes in local and organic foods, start in high school by working in the kitchen of a local restaurant. You might have to start as

Green culinary educators teach the philosophy and techniques of green dining. At a cooking camp in Arizona, children are taught to cook with organic vegetables.

a dishwasher, busboy, or prep cook, but you will be introduced to the fast-paced world of restaurant cooking. And you'll gain valuable insight into the skills and temperament needed to succeed. An advanced culinary degree, such as a certified chef de cuisine designation, is required for you to become a chef. You might earn a second degree in "green cuisine." A candidate for a chef position must have prior experience as a chef, cook, or caterer. The average hourly wage is $14 to $19. The average salary is $45,000 to $60,000. But being a chef has huge income potential, up to six figures, depending on the prominence of the establishment where you work.

Green Culinary Educator

A culinary educator may teach green culinary philosophies and techniques at a high school or college or in a continuing education class for working professionals. The green revolution makes this job more important than ever. The green culinary educator is on the frontlines, spreading the word about how to grow and obtain organic, Earth-friendly, local food and how to prepare it to keep people coming back. The culinary educator may be hired by environmentally conscious grocery and health food stores to help employees and customers wade through the confusing labels: "organic," "cage-free," "hormone-free," "antibiotic-free," "grass-fed," and so on.

To become a green culinary educator, you must have been an executive chef for at least two years. A culinary educator can work as a consultant, charging an average of $50 an hour. Or a culinary educator may receive a salary that starts at $50,000.

Chapter Seven

Jobs in Green Communications and Public Relations

One way to increase society's commitment to green and sustainable practices, including the conservation of popular tourist areas and unique communities, is by getting the word out—to consumers, government officials, and community activists. Education, information, and publicity are essential to make people aware of a problem,

In order to get travelers to come to this Costa Rican island, once home to a prison, the government of Costa Rica will need to get the message out to the world that the old jail has been converted to a resort and is now a very desirable destination.

motivate them to seek a solution, and energize them to see it through to its successful conclusion.

Community Conservation Specialist

A community conservation specialist works with local professionals, from forestry specialists to social scientists and business leaders, to develop protected natural areas for tourism. The job requires travel to locations where the conservation projects are being done. Community conservation specialists must identify, establish, and sustain protected areas and their buffer zones, and give equal consideration to landscape protection, cultural understanding, and sustainable livelihoods for local communities.

A master's degree in anthropology, human ecology, sustainable development, geography, or natural resource management with a social science emphasis is required for this position. Several years of work experience in community-based conservation, environmental conservation, or sustainable development are also recommended. Depending on location, a foreign language may be useful. Salaries range from $45,000 to $55,000.

Green Marketing and Public Relations Consultant

Tourist destinations need good publicity. Just as tourists need to do their homework to make good environmental choices, eco-destinations need to reach out to their potential customers. This is where marketers and public relations (PR) consultants come in.

Colorful, eye-catching streamers are part of this public relations director's plan for drawing people to an event celebrating sustainable seafood.

Marketers and PR consultants can be very important to a successful sustainable tourism program. They develop print, television, and radio advertisements in an attempt to lure tourists to their destinations. They issue press releases about new developments or attractions regarding the destination. They may be hired by local business bureaus to strategize about how to get tourists to come to their city or town. Other employers might be state and federal commissions that want them to assess how ready a place is for tourists. "Consultants bring a fresh pair of eyes," says Bruce Dickson of Tourism Development Solutions. The green consultant can structure tourist information to make it more appealing to travelers. "[Tourists] don't want to know everything," says Dickson. "An example of being strategic is to give them just what they need to know. What's the best trail? Where's the best restaurant?"

For marketing jobs, some employers prefer candidates with a degree in business administration with an emphasis on marketing. For PR jobs, a degree in public relations, with courses in public affairs, public speaking, and technical writing, is a plus. You can complete the package by doing one or two environmental internships while you are still in school. For both types of jobs, a background in business with additional training in green tourism and hospitality is helpful.

If you work for a marketing or public relations firm, you will probably earn from $50,000 to $75,000 a year, with benefits. Certification can garner an even higher salary. PR professionals can get certified by taking an exam offered by the Public Relations Society of America. A freelance PR consultant can earn from $100 to $200 an hour, depending on his or her experience and reputation.

Environmental Advocate

Advocacy is a large part of the current environmental movement. An advocate brings issues and solutions to the attention of lawmakers. This can be critical to many sectors of green travel and tourism. For instance, the American Recovery and Reinvestment Act of 2009 (popularly known as the "stimulus package") included hundreds of billions of dollars for new job creation. Many organizations wanted to make sure that some of this money went to green building projects. One of these was the American Institute of Architects, which hoped to see the money go to sustainable communities, new incentives for affordable housing, historic preservation, and conservation projects. These efforts would help the environment, while also employing thousands of architects, planners, and builders. Environmental advocates argue the case for green projects to lawmakers and local zoning boards.

Different kinds of companies, from green PR firms to green architectural firms, have at least one environmental advocate on staff. An environmental advocate's job can include any or all of the following: advocacy, policy development, campaign strategy, media outreach, and organization building (for example, writing grant proposals and meeting with large donors). In addition, an environmental advocate will probably mobilize and coordinate a pool of volunteers, who can help collect signatures for petitions or write to Congress on specific issues.

The ideal education for this position would include a bachelor's degree in environmental science, followed by a master's degree in environmental education and advocacy or in international development studies. Strong interpersonal

International Development Studies

At many universities, it's becoming popular to major in international development studies. This field of study focuses on the links between social, political, cultural, and economic factors in poor countries and the poor regions of wealthy countries. This would be a useful degree for any green travel and tourism professional. Relevant topics in this course of study include:

- History of international economic expansion
- Population resource issues
- Survival of indigenous societies
- Relationships between development and underdevelopment
- Cultural identity and cultural production

These programs are available at the University of California–Los Angeles, Ohio University, University of Montana, and McGill University in Canada, among others.

and writing skills are a must. Advocacy internships are a good way to gain experience. Such an internship can involve doing research and preparing reports, attending meetings with lawmakers, monitoring the progress of legislation in Congress, and preparing press releases. It can be exciting to take an active part in getting an environmental bill passed in Congress or to help a state pass strict renewable energy standards, all while getting your foot in the door. Internships can be paid or unpaid, and starting salaries for an environmental advocate range from $30,000 to $40,000.

The Wilds is a conservation, education, and research center for endangered species that live in the grasslands of southwest Ohio. Dr. Danny Ingold, a biology professor, hopes that his research on birds will increase ecotourism to the area.

Conservation Educator/
Natural Resource Educator

A conservation educator can work for government groups or private companies. He or she ensures that all parties in a conservation project understand what the goals are and how to reach them. For instance, in 2002, the Southwest Florida Water Management District began a project to promote water conservation in hotels and motels district-wide. This program is called Water C.H.A.M.P. (Conservation Hotel and Motel Program). One of the activities is to provide educational workshops for participating hotels and motels on how to save water. In 2008, some 420 properties participated in Water C.H.A.M.P., saving 177 million gallons (more than 670 million liters) of water.

Other employers for conservation educators include zoos, zoo associations, aquariums, parks, eco-parks, wildlife sanctuaries, and other nature recreation centers. If a conservation educator is working for a nonprofit nature education center, he or she will be responsible for conducting programs for schoolchildren and groups with special needs, and planning summer camps and special events.

A natural resource educator should have a background in the sciences—a bachelor's or master's degree in biology, zoology, botany, wildlife, fisheries, plant management, or related subjects. The area of specialty will influence where he or she ends up working. Salaries vary depending on the employer. Private firms pay better, approximately $30,000 to $50,000 a year. With a nonprofit employer, the job pays a "living allowance" (for instance $6,000 for nine hundred hours of work), plus benefits and bonuses.

Glossary

accreditation The act of granting official recognition; official certification that set standards have been met.

botany The branch of biology that studies plants and their interactions with the environment.

boycott To refuse to attend or do business with a company, usually in protest of its policies or practices.

carbon monoxide An odorless, poisonous gas that is produced by the burning of fuels, as in a car engine.

carbon offset A measurement of carbon reduction activities. It represents a reduction in greenhouse gas emissions. One carbon offset represents the reduction of one metric ton of carbon dioxide.

chlorofluorocarbons A group of chemical compounds, including methane or ethane, used in fire extinguishers and refrigerators. They have negative effects on the environment, including depleting the ozone layer.

ecotourism Tourism to exotic ecosystems to observe threatened wildlife, be educated about challenges to it, and help conserve and preserve it.

expedition A journey organized for a particular purpose.

fauna The animal life of a particular region.

flora The plant life of a particular region.

fossil fuel Fuel derived from the remains of organisms. Fossil fuels include coal, gas, and oil. They are a nonrenewable source of energy.

global warming An increase in the average temperature of Earth's atmosphere, largely caused by the buildup of greenhouse gases.

greenhouse gas A gas that traps heat in Earth's atmosphere, leading to global warming and climate change. This is known as the greenhouse effect.

indigenous A person, plant, or animal that is native to the particular area in which it lives.

nitrous oxide A chemical compound that is a major greenhouse gas. It is also known as "laughing gas" and is used in surgery and dentistry to block pain.

organic food Food that is grown without chemical fertilizers, pesticides, or hormones.

outfitter A shop that provides clothing and equipment for outdoor activities.

sealant A substance, such as wax, plastic, or silicone, that is used to seal something (to create a tight closure against the passage of air or water).

tsunami A huge destructive wave, sometimes caused by an earthquake on the ocean floor.

For More Information

International Guide Academy
P.O. Box 370190
Denver, CO 80237
(303) 780-0131
Web site: http://www.bepaidtotravel.com
The International Guide Academy is committed to pro-
    viding the tourism industry with the very best trained
    international tour managers and tour guides.

National Parks Conservation Association (NPCA)
1300 19th Street NW, Suite 300
Washington, DC 20036
(800) NAT-PARK (628-7275)
Web site: http://www.npca.org
The mission of this nonprofit education and advocacy
    group is to protect and enhance America's national
    parks system.

Natural Resources Defense Council (NRDC)
40 West 20th Street
New York, NY 10011
(212) 727-2700
Web site: http://www.nrdc.org/energy
The NRDC uses law, science, and the support of 1.2
    million members and online activists to protect wild-
    life and ensure a safe, healthy environment for all
    living things.

Nature Conservancy
4245 North Fairfax Drive, Suite 100
Arlington, VA 22203-1606
(703) 841-5300
Web site: http://www.nature.org
The Nature Conservancy identifies threats to marine
    life, freshwater ecosystems, forests, and protected
    areas, and then uses scientific approaches to
    save them.

Ontario Ecotourism Society
297 College Drive
Box 839
Haliburton, ON K0M 1S0
Canada
(705) 457-1680, ext. 6754
Web site: http://toes.ca
This Canadian nonprofit organization is devoted to
    the development and promotion of sustainable
    tourism practices. It supports ecotourism in the
    province of Ontario.

Outward Bound
Wilderness Expeditions
910 Jackson Street
Golden, CO 80401
(866) 467-7651
Web site: http://www.outwardbound.org
Outward Bound is a nonprofit educational organization
    that provides outdoor expeditions for groups of all
    kinds, including at-risk youth.

Rocky Mountain Institute
2317 Snowmass Creek Road
Snowmass, CO 81654
(970) 927-3851
Web site: http://www.rmi.org
This nonprofit organization fosters the efficient and
    restorative use of resources to make the world secure,
    just, prosperous, and life-sustaining.

Sustainable Buildings Industry Council (SBIC)
1112 16th Street NW, Suite 240
Washington, DC 20036
(202) 628-7400
Web site: http://www.sbicouncil.org
The SBIC champions sustainable buildings, which use
    renewable energy sources, conserve and protect
    water, use environmentally preferable products,
    and more.

Tour Guide Training Corporation of Canada
397 Front Street West, Suite #1908
Toronto, ON M5V 3S1
Canada
(647) 477 6470
Web site: http://www.tourmanagertraining.com
This Canadian school provides training for different types
    of tour guides.

Tourism Development Solutions (TCDS)
P.O. Box 19654
Portland, OR 97280

(503) 977-2772
Web site: http://www.developtourism.com
The TCDS provides clients with inventive solutions to their
   tourism strategy development, brand development,
   and niche tourism planning needs.

U.S. Green Building Council (USGBC)
2101 L Street NW, Suite 500
Washington, DC 20037
(800) 795-1747
Web site: http://www.usgbc.org
The USGBC seeks to expand green building practices,
   green education, and its LEED (Leadership in
   Energy and Environmental Design) Green Building
   Rating System.

Web Sites

Due to the changing nature of Internet links, Rosen
Publishing has developed an online list of Web sites
related to the subject of this book. This site is updated
regularly. Please use this link to access this list:

http://www.rosenlinks.com/gca/trav

For Further Reading

Burnett, Jim. *Hey Ranger!: True Tales of Humor and Misadventure from America's National Parks.* Lanham, MD: Taylor Trade Publishing, 2005.

Cassio, Jim, and Alice Rush. *Green Careers: Choosing Work for a Sustainable Future.* Gabriola Island, BC, Canada: New Society Publishers, 2009.

Chiras, Dan. *The New Ecological Home: A Complete Guide to Green Building Options.* White River Junction, VT: Chelsea Green Publishing Co., 2004.

Crosten, Glenn. *75 Green Businesses You Can Start to Make Money and Make a Difference.* Irvine, CA: Entrepreneur Media, Inc., 2008.

Environmental Careers Organization. *The ECO Guide to Careers That Make a Difference: Environmental Work for a Sustainable World.* Washington, DC: Island Press, 2004.

Freed, Eric Corey. *Green Building and Remodeling for Dummies.* Hoboken, NJ: Wiley Publishing, Inc., 2008.

Greenland, Paul R., and Annamarie L. Sheldon. *Career Opportunities in Conservation and the Environment.* New York, NY: Checkmark Books, 2007.

Hunter, Malcolm L., Jr., et al. *Saving the Earth as a Career: Advice on Becoming a Conservation Professional.* Malden, MA: Blackwell Publishing, 2007.

Inskipp, Carol. *Travel and Tourism* (The Global Village). North Mankato, MN: Smart Apple Media, 2008.

Llewellyn, A. Bronwyn, James P. Hendrix, and K. C. Golden. *Green Jobs: A Guide to Eco-Friendly Employment*. Avon, MS: Adams Media Corp., 2008.

Parks, Peggy J. *Ecotourism* (Our Environment). Farmington Hills, MI: KidHaven Press, 2005.

Rothschild, David. *Earth Matters*. New York, NY: DK Publishing, 2008.

Roza, Greg. *Reducing Your Carbon Footprint on Vacation* (Your Carbon Footprint). New York, NY: Rosen Central, 2008.

Sivertsen, Linda, and Tosh Sivertsen. *Generation Green: The Ultimate Teen Guide to Living an Eco-Friendly Life*. New York, NY: Simon Pulse, 2008.

Yudelson, Jerry. *Green Building A to Z: Understanding the Language of Green Building*. Gabriola Island, BC, Canada: New Society Publishers, 2007.

Yudelson, Jerry. *The Green Building Revolution*. Washington, DC: Island Press, 2007.

Bibliography

Brown, Jane Roy. "Looking to Share, Not Spoil, Riches." *Boston Globe*, December 28, 2008. Retrieved January 2009 (http://www.boston.com/news/local/vermont/articles/2008/12/28/looking_to_share_not_spoil_riches/?page=1).

Chon, Kaye Sung, and Simon Hudson. *Sport and Adventure Tourism*. New York, NY: Routledge, 2002.

Environmental Research Web. "Continental Airlines Flight Demonstrates Use of Sustainable Biofuels as Energy Source for Jet Travel." January 28, 2009. Retrieved February 2009 (http://environmentalresearchweb.org/cws/article/yournews/37558).

Federal Aviation Administration. "Aviation and Emissions: A Primer." Office of Environment and Energy, January 2005. Retrieved February 2009 (http://www.faa.gov/regulations_policies/policy_guidance/envir_policy/media/aeprimer.pdf).

Harris, Rob, Peter Williams, and Tony Griffin. *Sustainable Tourism: A Global Perspective*. Woburn, MA: Butterworth-Heinemann, 2002.

Henderson, Mark. "Green Tourism Tramples Over Wildlife Health." *Times* (UK), March 4, 2004. Retrieved January 2009 (http://www.timesonline.co.uk/tol/travel/holiday_type/green_travel/article515745.ece).

Herremans, Irene M. *Cases in Sustainable Tourism: An Experiential Approach to Making Decisions*. New York, NY: Haworth Hospitality Press, 2006.

Jamieson, Walter, ed. *Community Destination Management in Developing Countries*. New York, NY: Haworth Hospitality Press, 2006.

Kanter, James. "How Do You Measure Green Tourism?" *New York Times*, October 6, 2008. Retrieved February 2009 (http://greeninc.blogs.nytimes.com/2008/10/06/is-there-any-such-thing-as-green-tourism).

Lydersen, Kari. "Scientists: Pace of Climate Change Exceeds Estimates." *Washington Post*, February 15, 2009. Retrieved February 2009 (http://www.washingtonpost.com/wp-dyn/content/article/2009/02/14/AR2009021401757.html?hpid=topnews).

Sharma, K. K. *World Tourism Today*. New Delhi, India: Sarup & Sons, 2004.

Stewart, John. "Flying Shame." *Red Pepper*. Retrieved January 2009 (http://www.redpepper.org.uk/article319.html).

Swarbrooke, John, et al. *Adventure Tourism: The New Frontier*. Woburn, MA: Butterworth-Heinemann, 2003.

Tefler, David J., and Richard Sharpley. *Tourism and Development in the Developing World*. New York, NY: Routledge, 2008.

Wadhwan, Junie. "Ecotourism: Hope and Reality." *People and Planet*, August 5, 2008. Retrieved February 2009 (http://www.peopleandplanet.net/doc.php?id=1143&section=10).

Index

About the Author

Jacqueline Ching wrote for *Newsweek* and the *Seattle Times* before editing books. She is the author of many books for teens and young adults, including *Adventure Racing and Camping: Have Fun, Be Smart* for Rosen Publishing. She also writes articles about travel and environmental issues for Suite101.com.

Photo Credits

Cover (back, front) © www.istockphoto.com/Martin Strmko; p. 1 (right) © www.istockphoto.com/phil berry; p. 1 (left) © www. istockphoto.com/Jan Rihak; pp. 4–5 © www.istockphoto.com/ Ogen Perry; pp. 8, 44 © Salem Krieger/ZUMA Press; p. 11 © Joe McDonald/Visuals Unlimited, Inc.; p. 13 © Richard Hermann/ Visuals Unlimited, Inc.; p. 16 John & Lisa Merrill/The Image Bank/ Getty Images; p. 17 © Sarah-Maria Vischer/The Image Works; p. 21 Jewel Samad/AFP/Getty Images; p. 24 Banyu Sakti/AFP/ Getty Images; p. 26 © Paul Kennedy/Lonely Planet Images; p. 31 Daniel Garcia/AFP/Getty Images; p. 34 David McLain/Aurora/ Getty Images; p. 36 Mario Tama/Getty Images; p. 39 Bob Bird/Getty Images; p. 41 © John Gastaldo/San Diego Union-Tribune/ZUMA Press; p. 42 USDA Forest Service; pp. 47, 53, 57, 58, 60, 66 © AP Images; p. 49 © www.istockphoto.com/David Cannings-Bushell; p. 62 © Scott Keeler/St. Petersburg Times/ ZUMA Press.

Designer: Sam Zavieh; Photo Researcher: Amy Feinberg